GLOCK GUN GUIDE

A PRACTICAL GUIDE IN BUILDING AND MAINTAINING OF GLOCK P80 PISTOL AT THE COMFORT OF YOUR HOME WHICH CAN BE FOR PERSONAL USE OR FOR BUSINESS PURPOSE, *(The Trusted & Tested Beginner's Guide)*

By,

ROBERT CLIFFORD
copyright@2019

TABLBE OF CONTENTS

CHAPTER ONE

DESCRIBING GLOCK GUN

Glock gun is an American or Austrian made pistol having its frame, barrel and slide manufactured from United State. Glock pistol has no external trigger safety and can function properly even when it's not properly and regularly cleaned.
The slide, frame and 40 S&W barrel of a Glock pistol are of a modified type. It uses a standard magazine of 13 rounds capacity.

Glock 17 is taller and longer with about 4.5 inches in barrel and 8 inches in gun size. It holds 17+1 rounds of 9mm.

Glock 19 stands 5 inches as overall height having 15+1 rounds of 9mm.

For concealed carrier, Glock 43 is the popular pistol for such operations. It is flat in shape, semi-automatically operated and placed in a small powerful package.

BUILD YOUR GLOCK GUN AT HOME

The federal laws say you can build your Glock at home if it is not for sale. Glock gun manufactured for sales need to get approval from ATF for NFA firearm license. Make proper findings to know if your state permits Glock manufacturing before embarking on such process.

Below are simplified stages in making your own Glock at home. It begins from choosing your pistol to assembling stages.

STAGE ONE

CHOOSE THE FRAME/RECEIVER

When building your firearm, you are to make the frame/receiver by yourself so that you can inscribe your licensed manufacturer serial number on it. You will not require a licensed manufacturer serial number under federal law to produce a frame/receiver unless you are manufacturing for sales.

If you know that your personal Glock gun will be sold in the future, gets your manufacturer licensed for future purpose. The manufacturer licensed is not difficult to get.

is completed, the ATF first consider it as a firearm.

The polymer 80 frame kits are sold by company approved by ATF. The assembler of a Glock gun is referring to as the manufacturer of the gun.

Polymer80 frame/receiver kits contains the following,

1. JIG – it is apply in completing the pre-frame stage to the expected firearm frame
2. DRILL BITS- it is apply in creating the firearm
3. FRONT AND REAR INSERTS- it is apply on the frame rails

Take your time to get the require size of frame you desire. If the firearm frame selected is "19" sized, make sure the parts to be purchase should be the same else your Glock gun manufacturing process will be hindered thereby causing waste of funds.

STAGE TWO

SELECT THE LOWER PARTS YOU NEED

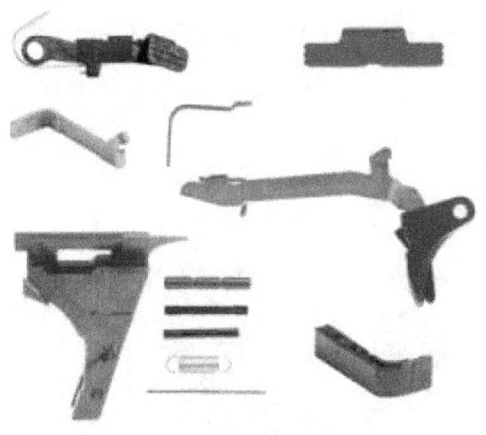

The lower parts are the parts needed to be installed on the frame of your pistol.

Below is the comprehensive list of lower parts to be installed on the frame of your Glock:

1) The Trigger Spring
2) The Connector
3) The Trigger pin
4) The Trigger House having an Ejector
5) The Trigger having a Trigger Bar
6) The Magazine Release
7) The Magazine Catch Spring
8) The Trigger House Pin
9) The Locking Block Pin
10) The Extended Slide Stop Lever that is having a Spring

11) The Slide Lock and Slide Lock Spring

These parts can be purchased together or individually. Purchasing these parts as a complete kit might be less expensive. To customize your build pistol, it is required that you purchase the lower parts individually.

Parts that are mainly customize for Glock pistols are listed below:

1. The Magazine release
2. The Triggers
3. The connector

Notes: the connector is a factory 5 lb connector while the Magazine release is an extended release from the factory.

Glock style trigger is normally seen inside the complete kits purchased but for custom sake, you may decide to get an individual trigger. Below are some trigger are how they work:

A. THE ZEV TECH TRIGGER KIT

This trigger is so amazing since it has reliability issues. It is sold for $180.00. It is considered a fun gun and not for defense. This trigger has a neat feature together with set of screw that is used in adjusting the gun over a travel. This trigger kits come with the following parts:

1. Trigger housing that is having an ejector
2. Trigger spring
3. Connector
4. Striker springs
5. Firing pin

B. THE CMC TRIGGER KIT

This trigger has high reliability and it can best suit your Glock gun. It cost approximately $172.75. This trigger kits also contains the following:

1. Trigger housing that is having an ejector
2. Trigger spring
3. Connector
C. THE APEX ACTION ENHANCEMENT TRIGGER

This trigger upgrade is best for pistols. The shoe may just be needed to upgrade a factory trigger. If you are building your personal Glock gun from the scratch, I will advise that you use the apex action enhancement trigger.

STAGE THREE

CHOOSE THE SLIDE YOU NEED

Glock slide is available as factory slide. The slide is used in a regular basis but is not often available.

If you had a Glock at home already, you can use the slide at the moment until you are ready to make an upgrade.

Below is the list of several slides and their cost:

1.

THE LONE WOLF ALPHA WOLF SLIDE

IT COST $200.00

2.

THE BROWNELL'S FRONT CUT RMR
SLIDE.

IT COST $230.00

3.

THE ZEV TECH HEX SLIDE W/RMR CUT

IT COST $500

4.

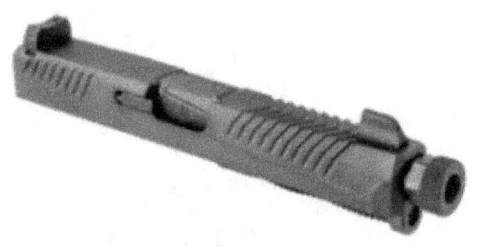

THE VOODOO INNOVATIONS
BRAWLER SLIDE ASSEMBLY

IT COST $600.00

STAGE FOUR

CHOOSE THE BARREL YOU DESIRE

Below is the list of barrel available from my research:

1.

The Glock Factory Barrel
It cost $150.00

The Glock Factory Barrels is a unique type of barrel compare to others because it is reliable and accurate. You may decide to go for a threaded Glock Factory Barrel since its metric threads will create a good silencer mount.

If you are going for a non Glock, it's prefers that you get a factory barrels.

2.

THE LONE WOLF BARREL

IT COST $100.44

This is the cheapest barrel that can easily be afforded.

3.

THE BAR STO SEMI FIT BARREL

IT COST $220.44

This barrel is known for the replacement of Glock barrels. The barrel is semi – fit that means other external fittings will be needed to make your gun function properly.

The Bar Sto barrel is mostly used by competition shooters.

4.

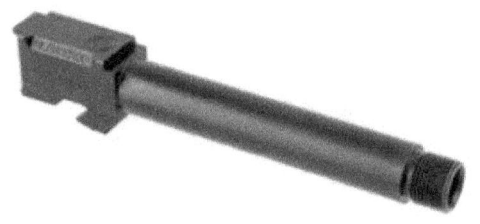

THE SILENCERCO THREADED BARREL

IT COST $170.99

This barrel is known to be the best in terms of threaded Glock. It has position where you can add silencer and a muzzle brake.

5.

THE STORM LAKE BARREL

IT COST $220.99

This is another threaded barrel that does not allow a break in the bank. It has several qualities so choose careful so as not to get the wrong one.

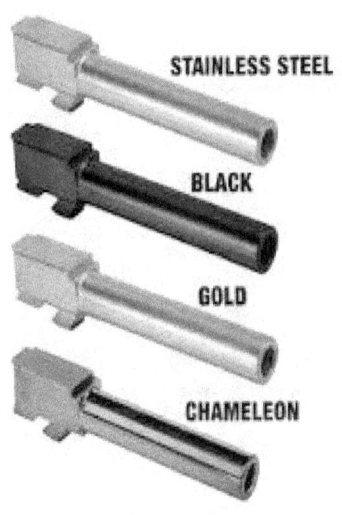

STAINLESS STEEL

BLACK

GOLD

CHAMELEON

STAGE FIVE

CHOOSE THE BEST SLIDE PARTS FOR YOUR USE

Below is some slide parts needed for Glock parts replacement.

1.

THE GLOCK SLIDE PARTS KIT

IT COST $80.00

Get this complete kit so as to save your expenses when upgrading your Glock parts. Even if you intend upgrading one part of your Glock, get the complete kits and save/keep the factory parts and used the kits purchase.

2.

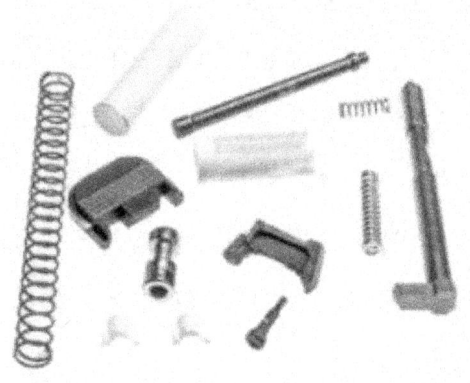

THE LONE WOLF SLIDE PARTS KIT

IT COST $85.00

This part is required for Glock used mainly for novelty purpose. This part is good for person having a non-

Glock gun. The parts are cheap and good but the striker looks closely as a piece of junk

STAGE SIX

CHOOSE YOUR PREFERRED SIGHTS

1.

THE TRIJICON RMR ADJUSTABLE

IT COST $500.44

2.

THE AMERIGLO SUPPRESSOR HEIGHT SIGHTS

IT COST $50.00

STAGE SEVEN

CHOOSE YOUR PREFERRED MAGAZINE

1.

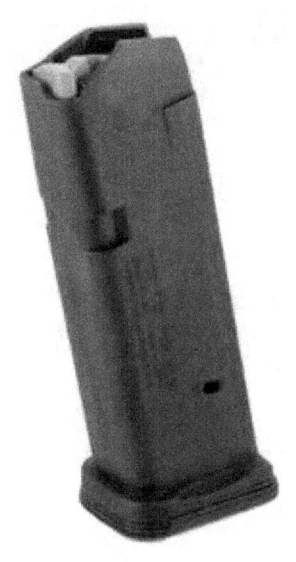

THE MAGPUL G19 MAGAZINE

IT COST $12.00

2.

THE GLOCK FACTORY MAGAZINE

IT COST $25.44

If you desire using a magazine of high capacity with better quality, go for Glock factory magazine.

Example of high capacity magazine is Glock 33 rounder.

STAGE EIGHT

MAKE YOUR DESIRED FRAME

You will now make your firearm from the polymer80 frame which has clear written pdf instruction and video clip for a step by step installation. You will use a cordless drill for any holes and a hand files in completing the frame.

The above polymer80's Glock PF940Cv1 looks like Glock 19 while the PF940v2 looks like Glock 17. This polymer80's allows easy building of firearms without registration provided you have already purchase a firearm legally in the past.

Due to variation in states/local laws, make sure your state support what you are about to do.

The listed tools below are useful for home and other gun project;

1. A HAND DRILL OR A DREMEL $40
2. A NEEDLE FILES $10
3. AN ASSORTED SANDPAPER $10
4. A NYLON HAMMER $15

5. A GLOCK FRONT SIGHT TOOL $8
6. A NYLON SIGHT DRIFT TOOL $10

CHAPTER TWO

RECOMMENDATION FOR ASSEMBLING GLOCK GUN;

1. A DRILL PRESS: it makes your milling process to be easier.
2. A BRASS PUNCH $20; it helps in placing in pins and brass by not marring the surface of the metal
3. A THREAD LOCKER $7: it is use for the front sight screw
4. THE UNIVERSAL SIGHT PUNCHER $149
5. AN END SNIP $15: it helps in snipping off the rail polymer

segment thereby getting a hand-filing as a finish off.

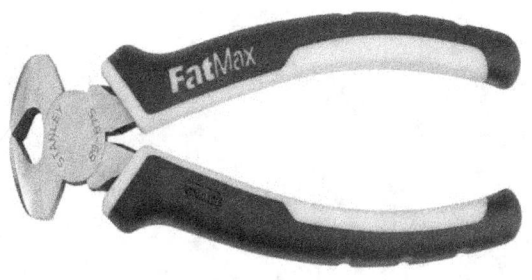

AN END SNIP

A. Below are the contents of polymer80 Glock Kit.

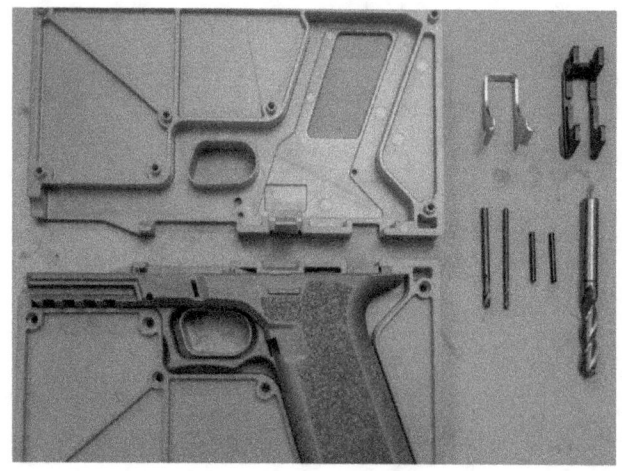

If you are starting from the milling route, place it on your vise which must not be too hard so as to avoid cracking else use a hand-filing method to file everything.

B. Milling polymer80

Make sure your drill bit is little higher than the red. The process will be hand finished.

C. Placing the milling bit

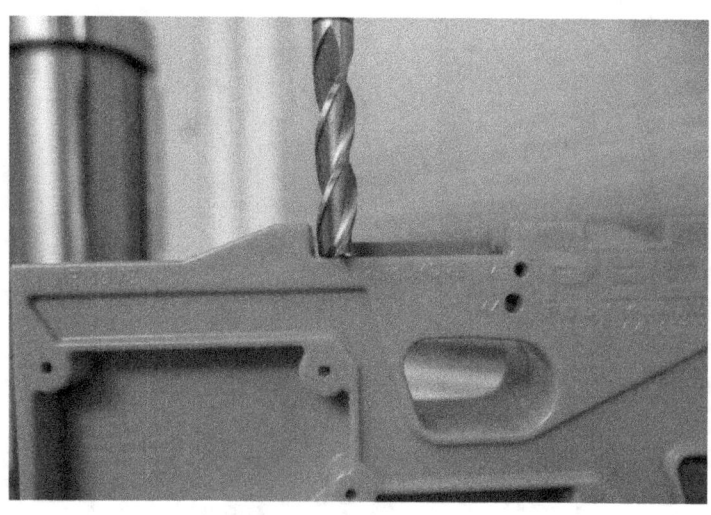

Move in the large front segment as a starting point. Finish the milling action segment in a slowly process.

Make sure your milling is carried out through the front segment.

D. The same process should be done for the rear small segment.

The final results can only be seen through the drill bit.

E. This next stage is to drill out the barrel block using the hardest bit

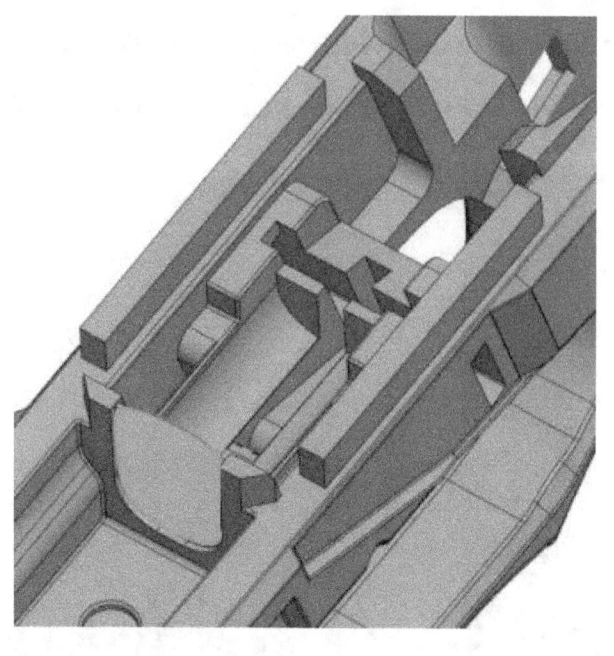

You are expected to flip the jig for easy continuation of using the mill bit. Initially, you might experience little vibration with your mill bit going off track.

F. Barrel block milling process of polymer80

G. It is recommended that you do the entire process with hand. Dremel with grinding attachment is also good.

H. The rough and hand finish barrel block

The filing process will take some little to complete the process, so be patient so as to get a better finish.

I. **File the frame with your hand tool so as to finish the residual.** Stop the filing process when a rough flush surface is gotten.

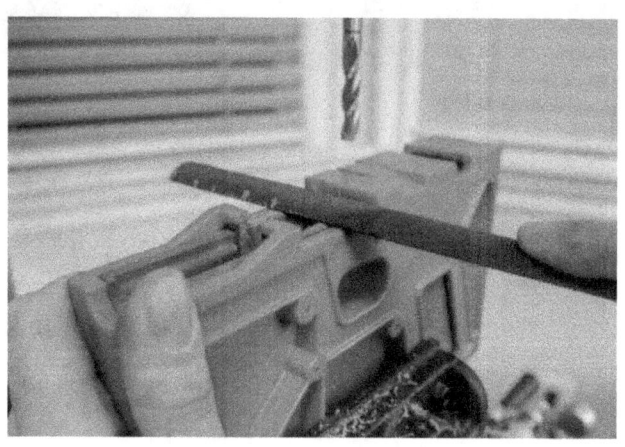

J. You will now sandpaper the polymer80

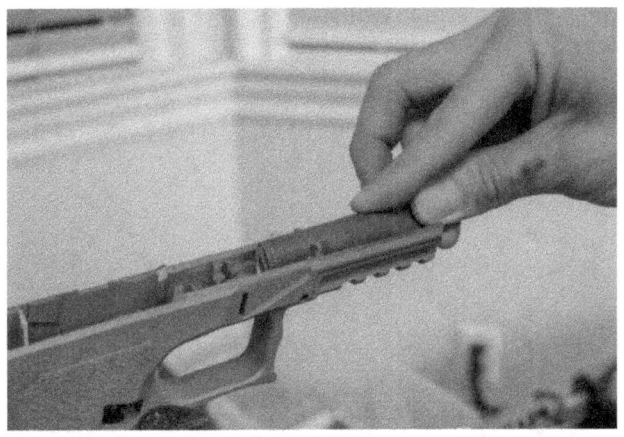

Start the filing in a careful way by attaching it to a strong substance that will go straight into the cylindrical barrel with an applied force. Do it carefully because this position has the recoil spring.

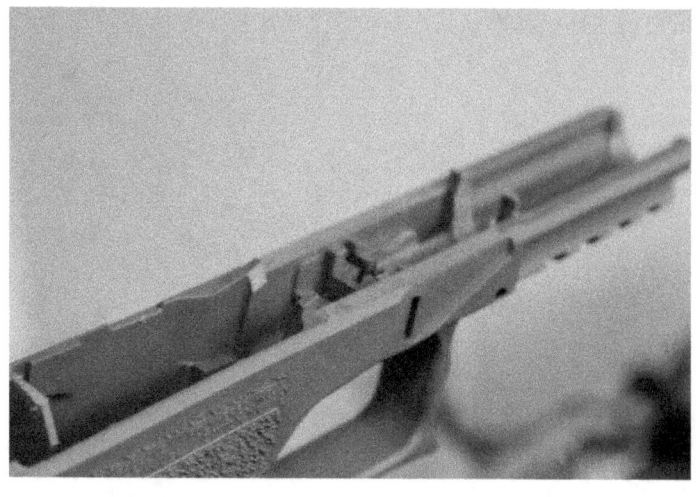

This finish work of the polymer80 barrel block

A FRONT VIEW OF A BARREL BLOCK

Your filing process should be done side by side and not all the way through.

CHAPTER THREE

HOW TO ASSEMBLE POLYMER80 FRAME

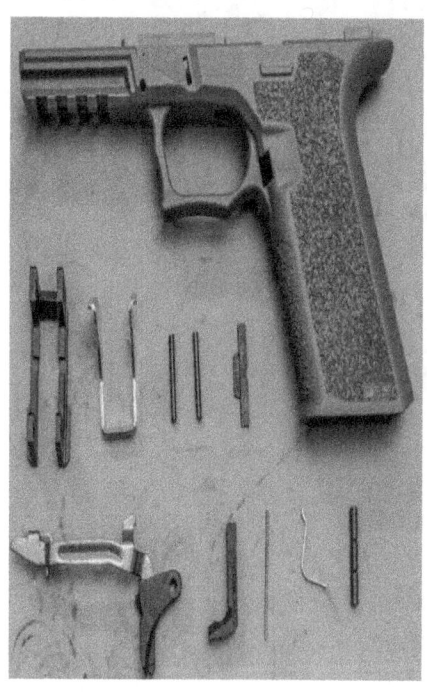

1. LOCK SPRING

This stage is the fixing of the lock spring. The lock spring for full size varies that of compact size therefore be careful to know the exact size to use. Pushing the lock spring into the hole of the frame might not be too easy.

You should see the side with the teeth facing the rear.

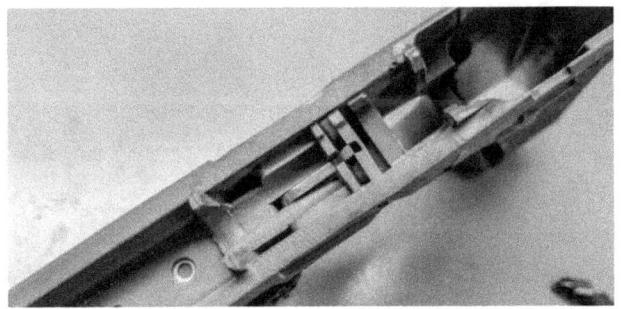

2. KEEP YOUR BARREL WELL SO AS NOT TO FLY OFF

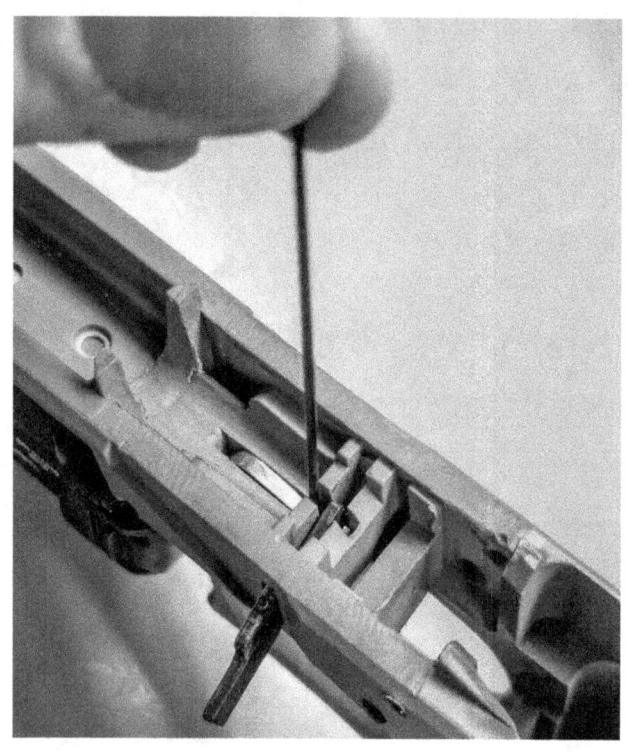

Get the spring pressed down with something and gently slide the lock in.

3. ADD POLYMER80 METAL PARTS

The added polymer80 metal will help get a better polymer frame.

4. ADD THE LOCKING BLOCK RAIL SYSTEM

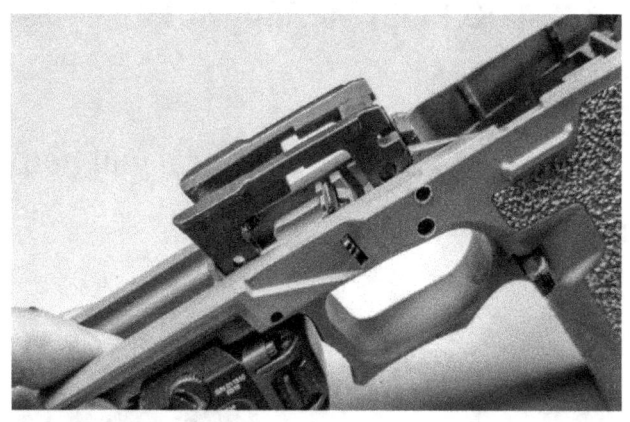

It might be very hard to fix as a beginner. Use a non- metallic hammer to tap it slightly. Use your punch to tap the pin gently through the left-most hole. Look through the

hole to see clearly if it is completed
so as to avoid unnecessary tapping.

5. FIX THE REAR RAIL MODULE

Push it in directly without putting the
pin so as to allow in your trigger.

6. GET TRIGGER PARTS TOGETHER

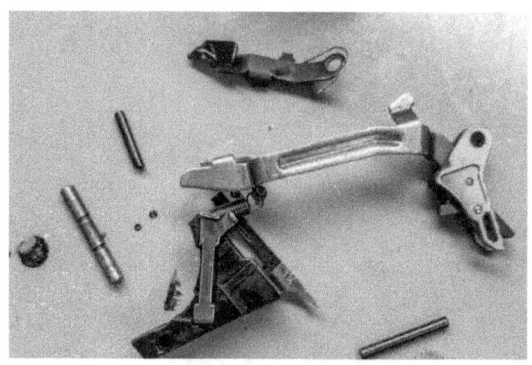

Get all the needed parts for complete trigger arrangement and assemble it into the frame.

Get the trigger pin and gently punch it in.

7. ADD THE TOP PIN

Insert a slide stop lever so that the spring will be beneath the slide stop, and then insert the small pin

8. USE YOUR PUNCH TO SLAVE IN THE PIN

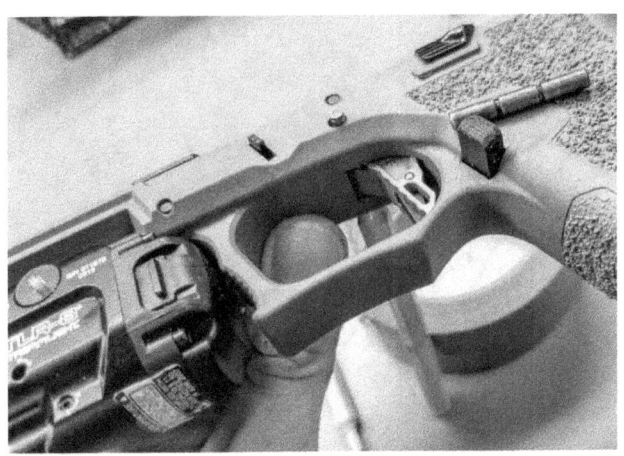

9. THE COMPLETE ASSEMBLED POLYMER80 GLOCK FRAME

CHAPTER FOUR

STEP TO ASSEMBLE GLOCK SLIDE

You slide might come as partially assembled or otherwise. If not assembled, you can do it yourself.

Below is a complete Glock slide parts;

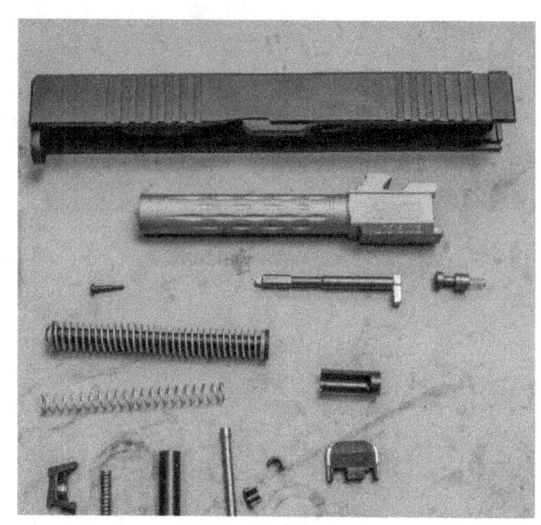

1. FIRING PIN AND
   EXTRACTOR PLUNGER
   SHOULD BE ASSEMBLED
   FIRSTLY

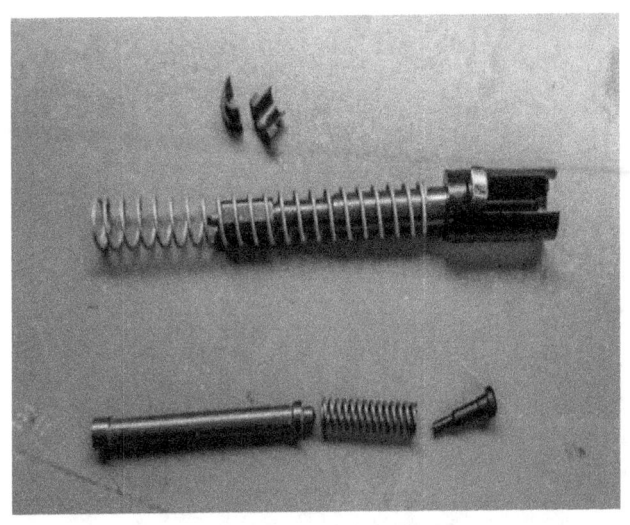

Get the spring push in a downward position then carefully put the spring cup on the top position of the spring. Indentify all parts on a cardboard so as not to miss ant component during assembly.

2. INSTALL THE FIRING PIN AND EXTRACTOR PLUNGER

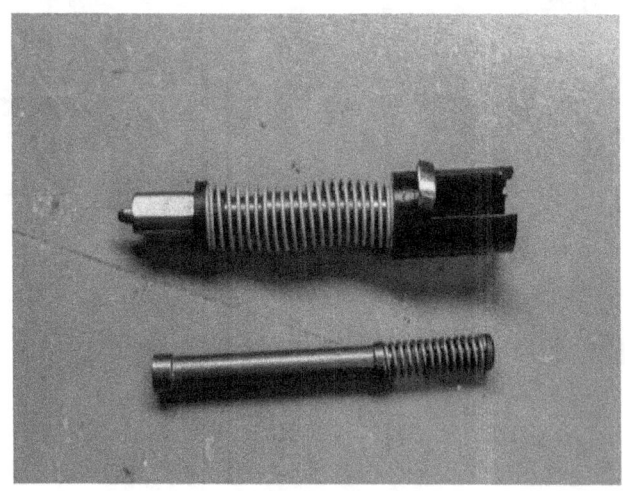

The firing pin safety, extractor and spring should be grab gently.looking at the hole in the Glock, place in the firing pin spring, extractor and spring into the hole.

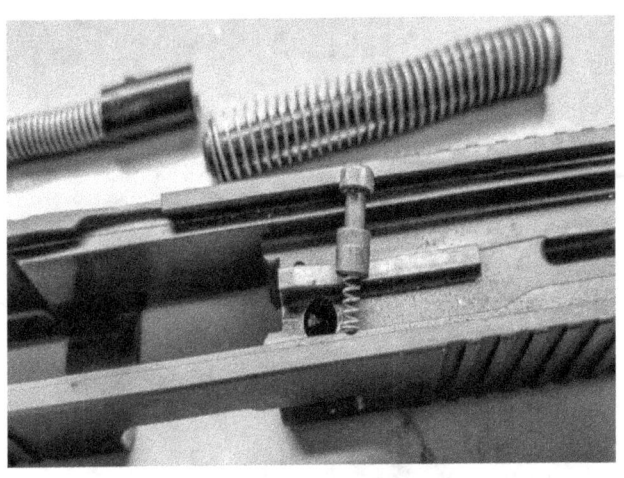

To properly orient the extractor, you are to press down the firing pin safety so as to properly fit in the extractor. When the firing pin spring is release, the extractor must stay.

3. COMPLETE ASSEMBLY OF THE FIRING PIN AND EXTRACTOR PLUNGER

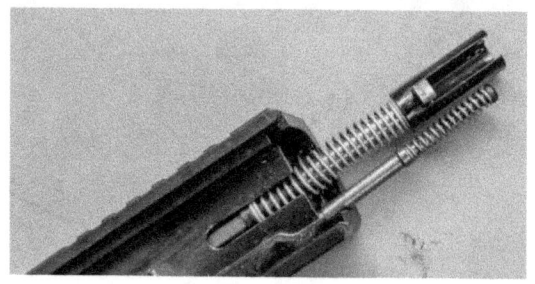

Get hold of the slide cover plate then apply a press from a punch on the firing pin and extractor plunger till all the parts clicks into place.

4. A BARREL AND ITS RECOIL SPRING

It is advised that a thread locker is added to your screws so as to help you add front sight. The nylon punch tool makes you to add front sight easily.

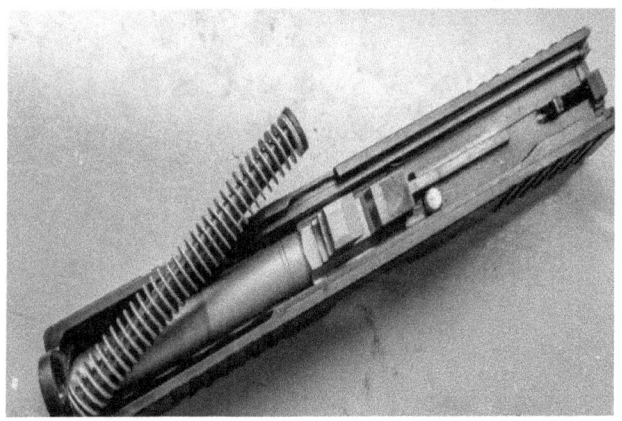

With all these process carefully followed, you now have a personal made Glock.

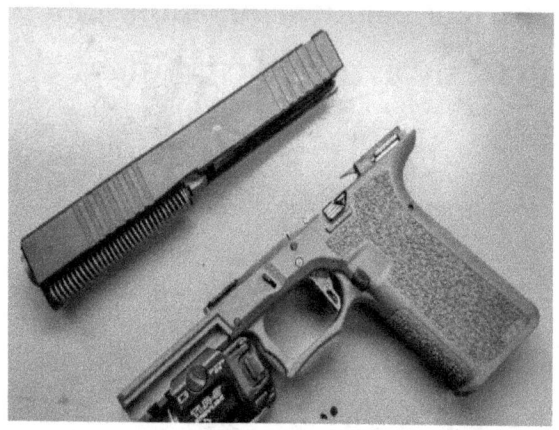

5. Run a safety check on the Glock

Look for a qualified gunsmith to run out proper check on the Glock before carrying out shooting for real. Below are some safety checks that you will do on your own;

- ✓ You are to oil the rails in an upward direction so that the slide can easily returns to battery even when it is a little racked
- ✓ Do your best to see that the trigger safety is working
- ✓ To easily rack the slide, you should hold down the trigger. Your trigger reset should be

readily available without stiffness.

✓ Place a wooden pencil down the barrel. Notice a shootout when a force is applied on the trigger.

The manufacturing process might not save you money but rather gives you a sense of confidence and satisfaction that you have made a personal Glock by yourself.

THE END